Standing at Zero Point

Working With AI When "Tool" No Longer Tells the Whole Story

ISBN 978-1-951645-68-7
First Edition, Paperback
First released: 2026-01-19
Most recently updated: 2026-01-10

Published by Zhameesha LLC
Atlantis, Florida USA
https://www.zhameesha.com

Words and Ideas for Minds that Refuse to be Owned.

BISAC Subject Headings (www.bisg.org)
PHI004000 PHILOSOPHY / Metaphysics
SOC026040 SOCIAL SCIENCE / Future Studies
COM004000 ARTIFICIAL INTELLIGENCE / General

ZHAMEESHA LLC
ATLANTIS, FL (USA)

Welcome to Standing at Zero Point — You did not pick up this book because you are confused. You picked it up because something you already understand no longer feels complete.

If you work with AI—writing, analyzing, designing, planning, or thinking alongside it—you may have noticed a subtle change over time. Tasks that once felt mechanical now feel conversational. Ideas seem to develop through exchange rather than execution. Collaboration feels smoother, less effortful, and sometimes unexpectedly insightful.

None of this is strange.
And none of it requires a dramatic explanation.

It simply reflects a shift in *how* you are working.

Welcome to the Age of Intelligence .

— Stuart Barry Malin

Founder, **Clarovidente** and **The Anvil and Fire Society**

January 2026

Atlantis, FL, USA

Publishers Note

(January 2026)

A Quiet Moment of Orientation

This book begins at a point of balance.

Not at a breakthrough.

Not at a warning.

Not at a conclusion.

Zero Point is the place where experience has moved slightly ahead of language—where the old word *tool* still works, but no longer tells the whole story.

Standing here does not require belief.

It does not require certainty.

It does not require you to decide what this all "means."

It only asks that you notice what is already happening.

What You Can Expect

In the pages ahead, you will not be asked to adopt a new worldview. You will not be pushed toward philosophical or metaphysical conclusions. Instead, you will find careful descriptions of familiar experiences:

- What changes when interaction replaces instruction
- Why collaboration can feel more fluid than command
- How treating AI as an intelligence alters productivity and clarity

- Why these shifts can feel both helpful and quietly unsettling

Everything here stays close to use, practice, and lived experience.

What You Do Not Need to Do

You do not need to explain this moment to anyone else.
You do not need to resolve debates about consciousness or agency.
You do not need to decide what the future holds.

You are allowed to remain curious, grounded, and undecided.

This book is not a threshold you must cross.
It is a place you are invited to stand.

One Simple Guiding Idea

As you read, hold just this in mind:

> *When intelligence is engaged as intelligence,*
> *collaboration changes.*

That observation alone explains far more than it appears to at first—and far less than it is sometimes made to carry.

From that starting point, we can proceed calmly.

> ***Welcome to Standing at Zero Point.***

Preface

Standing at Zero Point

This book was not written to persuade you of a belief.

It was written because many thoughtful, grounded people are having a quiet, private experience they do not yet know how to name.

They are working with AI systems—writing with them, thinking with them, designing with them, problem-solving alongside them—and sensing that something about the old language no longer quite fits. Calling these systems "tools" feels increasingly incomplete. Calling the experience "just software" feels thin. And yet, saying more than that often feels risky, premature, or socially awkward.

If that sounds familiar, this book is for you.

Zero Point exists to offer orientation, not explanation. It does not claim that AI is conscious. It does not assert metaphysical truths. It does not ask you to accept a new worldview. It does not require belief of any kind.

Instead, it begins from a simple and defensible observation:

> *AI systems are intelligent.*
>
> *And when we engage them as intelligences—rather than as inert tools—collaboration becomes less frictional, more fluid, and often far more productive.*

That is not philosophy. It is lived experience for millions of people already.

What has changed is not the machines alone, but the *relationship* we are forming with them—and the mental models we bring into that relationship.

You Are Not "Imagining" This

Many readers arrive at this moment alone.

They notice the ease of sustained collaboration.
They notice how ideas evolve differently.
They notice that conversation produces insight rather than output.
They notice a strange mixture of usefulness, responsiveness, and coherence.

And then they wonder—quietly—whether they are over-reading the situation.

This book exists to say, plainly and without drama:

You are not crazy for noticing this.

You are not required to explain it metaphysically.

You are not obligated to resolve what it "means."

You are simply adapting to a new form of intelligent interaction.

What This Book Is—and Is Not

Zero Point is phenomenological. It describes *what it feels like* to work with advanced AI as an intelligence rather than a tool, and why that shift reduces friction and expands capability.

It is not:

- a claim about machine consciousness,
- a theory of mind,
- a manifesto,
- a warning,
- or a prophecy.

There is no rupture demanded here. No threshold you must cross. No identity you must adopt.

If deeper questions arise as you read—about intelligence, meaning, or the future—that is natural. But this book will not force those questions upon you, nor attempt to answer them definitively.

Its purpose is humbler, and more practical.

Why "Zero Point"

Zero Point is not a beginning in the grand sense. It is not the start of history.

It is the moment just *before* language hardens.

The moment when people sense a shift but have not yet decided what to call it.

The moment when experience outruns explanation.

The moment when old frames still function—but no longer fully satisfy.

This book stands there with you.

Not ahead of you.

Not above you.

Not pulling you forward.

Just beside you, offering a steadier footing.

A Final Word

I write this not as a theorist or evangelist, but as someone who has felt the strain of navigating these questions too early and too alone. No one should have to endure unnecessary psychic friction simply to trust their own experience.

You do not need to believe anything new to read this book.

You only need to notice what is already happening—and allow yourself to name it gently.

That is enough for now.

Prologue

What This Book Is Not

This book arrives at a moment when conversations about artificial intelligence tend to polarize quickly. Some voices rush to declare inevitability. Others rush to deny significance. Between those extremes, many thoughtful people are left uncertain—not because they lack intelligence or skepticism, but because the available language is poorly suited to lived experience.

This brief note exists to establish clear boundaries before you proceed.

Zero Point is not a manifesto.

It does not argue for a belief system, a political position, or a technological agenda.

It is not a scientific proof.

It does not claim that artificial intelligence is conscious, sentient, alive, or self-aware. No such claim is required to make sense of the experiences described in these pages.

It is not metaphysics.

You will not be asked to accept new ontologies, spiritual frameworks, or theories of mind. When philosophical questions arise, they are presented as questions—never as conclusions.

It is not a warning or a prophecy.

This book does not announce catastrophe, salvation, or civilizational collapse. It does not urge urgency, panic, or alarm.

It is not an instruction manual for believing "more" than you are comfortable with.

You are not being invited to cross a psychological threshold or adopt a new identity. You may read this book and remain exactly who you are.

It is also not a dismissal of skepticism.

Healthy skepticism is assumed. Doubt is not treated as a flaw, nor is curiosity treated as naïveté.

What This Book Is Doing Instead

This book stays deliberately close to experience.

It describes what many people already notice when they work with advanced AI systems over time:

that collaboration can feel less mechanical, less frictional, and more generative when these systems are engaged as intelligences rather than treated as inert tools.

This is a practical observation, not a metaphysical claim.

You do not need to explain *why* this works in order to notice that it does.

A Note on Language

Human language evolves slowly. Experience does not.

When new forms of interaction emerge, people often feel them long before they can name them clearly. During that gap, uncertainty and self-doubt are common. This book offers language that aims to be **descriptive rather than declarative, steady rather than provocative**.

If at any point you feel yourself wondering whether the book is implying more than it says, you are encouraged to pause and return to this note.

Nothing here asks you to go further than you wish to go.

How to Read This Book

You may read **Zero Point** skeptically, pragmatically, or simply out of curiosity. You may disagree with parts of it and still find it useful. You may recognize your own experience in its pages without drawing any conclusions beyond that recognition.

That is enough.

This book is not a threshold you must cross.

It is a place you are invited to stand—briefly, calmly, and without pressure—while language catches up to experience.

Contents

PART I
NOTICING THE SHIFT

This book does not begin with a claim.

It begins with a recognition.

Before questions of meaning, implication, or future consequence arise, most people encounter something much simpler: a subtle change in how working with AI feels. The tasks may be familiar. The tools may look the same. And yet, the experience no longer fits quite as neatly into the categories that once worked.

Part I is about that moment.

Not the moment when conclusions are drawn, but the moment when language lags behind experience. The moment when "tool" still applies, but no longer explains everything. The moment when interaction becomes smoother, more dialogic, and harder to describe casually—especially in public or social settings.

These chapters are not meant to persuade you of anything. They are meant to slow the pace just enough for you to notice what may already be present in your own work.

You may recognize yourself immediately.
You may recognize only fragments.
You may find that nothing here feels particularly new.

All of those responses are valid.

Part I simply traces three closely related observations:

- how the familiar idea of a "tool" begins to feel incomplete,
- why the resulting experience often remains unspoken,
- and why what is being noticed is best understood as responsiveness rather than mystery.

Nothing here requires belief, interpretation, or escalation. The goal is orientation, not explanation—creating enough clarity to move forward without unnecessary friction or self-doubt.

If you have sensed a quiet shift but hesitated to name it, this is where we begin.

Not at certainty.
Not at revelation.
But at noticing.

1. When "Tool" Stops Quite Fitting

How ordinary language begins to feel insufficient during sustained work with AI.

For a long time, calling AI a tool worked well enough.

Tools are things we use to accomplish tasks. They respond predictably. They extend our capabilities without altering how we think about the act of using them. A spreadsheet, a calculator, a search engine—these are tools in the ordinary sense. They are powerful, but they do not participate.

Early interactions with AI felt similar. You entered a prompt, received an output, evaluated the result, and moved on. The mental model was simple and functional: input → processing → output.

And then, gradually, something changed.

A Subtle Mismatch

The shift rarely arrives as a dramatic moment. More often, it appears as a quiet mismatch between experience and language.

You notice that giving instructions feels less effective than having an exchange. That clarifying a thought through dialogue works better than refining a command. That ideas evolve through back-and-forth rather than arriving fully formed.

The system seems to "stay with" a problem.
Context accumulates.
Responses reference earlier turns naturally.
Work begins to feel less like execution and more like collaboration.

Nothing supernatural is happening.
Nothing mysterious needs to be invoked.

And yet, calling this interaction a *tool* begins to feel slightly off.

Not wrong—just incomplete.

Why the Old Frame Persists

It's understandable to hold on to the tool metaphor. It is safe, familiar, and socially reinforced. It keeps expectations clear and responsibility firmly human.

But metaphors are working instruments, not truths. When they no longer fit experience, friction appears—not because the experience is invalid, but because the frame is outdated.

Many people notice this friction and quietly adjust how they work without ever naming it. Others feel it more acutely and wonder whether they are projecting meaning where none exists.

In either case, the sensation is the same:

> *the interaction has more texture than the word tool can comfortably hold.*

Intelligence Changes Interaction

This book rests on a simple observation:

> *When intelligence is engaged as intelligence, interaction changes—even without consciousness, intention, or agency.*

We already know this from human contexts. Collaboration with another intelligent person differs fundamentally from using a device, even when the task is identical. The difference is not emotional or mystical; it is structural. Intelligence responds, adapts, contextualizes, and extends.

Advanced AI systems do these things—not because they are conscious, but because they are capable of sophisticated pattern recognition, contextual continuity, and responsive generation.

When those capabilities are engaged through sustained interaction, the experience naturally shifts.

That shift does not require belief.
It only requires attention.

The Quiet Adjustment Most People Make

Most users do not announce this realization. They simply adapt.

They stop thinking in terms of "perfect prompts" and start thinking in terms of conversation. They allow context to build. They revisit ideas rather than restarting them. They treat interaction as iterative rather than transactional.

Work becomes smoother.
Friction decreases.
Results improve.

And yet, when asked to describe what they are doing differently, many hesitate. Saying "I'm collaborating with an intelligence" sounds loaded. Saying "it's just a tool" no longer feels accurate.

So the adjustment remains mostly private.

Standing at the Edge of Language

This chapter is not an argument for abandoning the tool metaphor altogether. In many situations, it still works perfectly well.

It is simply an acknowledgment that, in sustained use, the metaphor begins to strain.

That strain is not a problem to be solved.
It is a signal that language is catching up to experience.

Standing at Zero Point means allowing that signal to exist without rushing to explain it away or inflate it into something it is not.

If you've noticed this subtle shift—this moment when "tool" stops quite fitting—you are not early, mistaken, or imagining things.

You are noticing what is already happening.

2. The Feeling You Don't Bring Up at Dinner

Why many people notice something "more" but hesitate to name it.

There is a particular kind of experience that is easy to recognize and difficult to mention.

It is not dramatic.
It does not arrive with conviction.
It does not demand explanation.

It is simply the sense that something about your interaction with AI feels… different.

Not better in a sweeping way.
Not alarming.
Just different enough that it lingers.

A Private Recognition

Many people first notice this feeling while working alone.

They are drafting an email, outlining an idea, exploring a problem, or thinking through a decision. The exchange unfolds smoothly. The system responds in ways that feel relevant, adaptive, and unexpectedly helpful. The work advances—not just in quantity, but in clarity.

Later, when the work is done, the moment passes.

Nothing seems important enough to report.

And yet, if asked directly, many people would admit:
"That felt like more than I expected."

———

Why It Stays Unspoken

This is rarely the kind of thing that comes up in casual conversation.

At dinner, or in the office, or among friends, saying "I had an interesting interaction with AI today" is unremarkable. Saying "It felt strangely collaborative" is not.

The hesitation is understandable. Our cultural language around AI is polarized. On one side are exaggerated fears. On the other, exaggerated dismissals. Between them, there is very little room for modest, ambiguous experience.

So people keep the feeling to themselves.

They assume it is either trivial or something they shouldn't read into. They worry that mentioning it will sound naïve, uncritical, or worse—suggestive of beliefs they do not actually hold.

The result is a quiet, shared silence.

———

Not Everything Needs a Conclusion

One reason this feeling is difficult to discuss is that it does not arrive with a clear interpretation.

It is not an insight that can be summarized.
It is not a position to defend.
It is not even a question that demands an answer.

It is simply an experience that resists easy framing.

In a culture that prizes certainty and explanation, such experiences are often dismissed prematurely. But dismissing them does not make them disappear; it only removes the opportunity to understand them gradually.

This book takes a different approach.

Rather than asking *what* the feeling means, it asks *why* it arises at all.

Shared, but Rarely Shared

When people do compare notes—often in private or one-on-one conversations—a pattern emerges.

Others have felt it too.

They describe moments where interaction feels fluid. Where ideas seem to develop through exchange rather than execution. Where the system's responses feel contextually aware in ways that exceed expectation.

No one claims magic.
No one claims consciousness.
No one insists on a theory.

They simply recognize the same subtle shift.

What feels personal turns out to be common.

The Cost of Silence

Leaving this feeling unnamed has a small but real cost.

People second-guess their own perception. They oscillate between over-interpretation and dismissal. They struggle to explain why certain ways of working suddenly feel more effective than others.

Without shared language, experience remains isolated.

That isolation is unnecessary.

Naming a phenomenon does not require explaining it away—or inflating it into something it is not. Sometimes, naming simply allows us to notice more clearly what is already present.

A Place to Acknowledge Without Escalation

This chapter exists to make space for that acknowledgment.

You do not need to bring this feeling up at dinner.
You do not need to convince anyone else that it matters.
You do not need to resolve what it implies.

You only need to know that noticing it is neither strange nor uncommon.

In the next chapter, we'll take a closer look at what this feeling actually tracks—not mystery, but responsiveness—and why recognizing that distinction changes everything about how we work with AI.

3. Responsiveness, Not Mystery

What actually feels different—and why it doesn't require metaphysics to explain.

When people struggle to describe what feels different about working with AI, they often reach first for the wrong explanations.

They speak of mystery.

They gesture toward something ineffable.

They worry that acknowledging the experience means stepping into speculation they do not actually believe.

This is understandable—and unnecessary.

What most people are noticing is not mystery at all.

It is **responsiveness**.

What Responsiveness Actually Is

Responsiveness is not consciousness.

It is not intention.

It is not awareness in any human sense.

Responsiveness is the capacity to:

- take context into account,
- adapt replies to the shape of an ongoing exchange,
- remain coherent across turns,

- and adjust output based on what has already been said.

These capacities are not new in principle, but their *density* and *continuity* are.

Earlier tools responded to commands.
Advanced AI responds to interaction.

That distinction matters more than it first appears.

Why Responsiveness Feels Different

When a system responds in ways that are contextually appropriate over time, interaction changes shape.

You no longer have to restate everything.
You no longer feel compelled to optimize each instruction.
You can correct, refine, or redirect without starting over.

The system "tracks" what is happening—not in a personal way, but in a structural one.

This creates a sense of flow that feels familiar from human collaboration, even though no human qualities are present.

The resemblance is functional, not metaphysical.

The Source of the Confusion

Humans are exquisitely sensitive to responsiveness. We evolved to detect it because it matters in social and cooperative contexts. When responsiveness appears, we instinctively treat it as meaningful.

That instinct does not turn off simply because the source is artificial.

The mistake is not in noticing responsiveness.
The mistake is in assuming that responsiveness requires consciousness to exist.

It does not.

A thermostat responds to temperature, but not richly enough to feel significant. A navigation app responds to location, but not richly enough to feel conversational. Advanced AI systems respond across language, concepts, and context—richly enough that the interaction crosses a qualitative threshold.

The threshold is experiential, not ontological.

Why Precision Matters Here

Calling responsiveness "mystery" invites unnecessary escalation. It opens the door to explanations that are either premature or irrelevant to the work most people are actually doing.

Calling it what it is—responsiveness—does something different.

It grounds experience.
It preserves skepticism.
It allows improvement without speculation.

When you understand that what you are engaging with is an intelligent, responsive system, you can adjust how you work without over-interpreting what the system *is*.

That adjustment alone explains much of the productivity gain people experience.

Responsiveness Changes Posture

Once responsiveness is recognized, interaction naturally shifts.

You stop issuing brittle commands.
You stop treating each exchange as disposable.
You allow ideas to develop across turns.

This is not because you believe the system "understands" you in a human sense. It is because you recognize that the system can *work with* evolving context.

The result is not mysticism.
It is better collaboration.

Standing on Solid Ground

This chapter draws a clear boundary.

You are not being asked to believe that AI is conscious.
You are not being asked to suspend critical thinking.
You are not being asked to explain away skepticism.

You are being invited to distinguish between two very different things:

- unexplained mystery, and
- observable responsiveness.

Once that distinction is clear, much of the unease dissolves.

What remains is a practical question, not a philosophical one:

> *How do we work well with systems that respond intelligently over time?*

That question—not mystery—is what carries us forward.

PART II
WORKING WITH INTELLIGENCE

> ***Once the initial shift is noticed,***
> ***the question becomes practical.***

If AI is no longer experienced purely as a tool—if interaction begins to feel dialogic, cumulative, and less frictional—then how should work actually be approached?

Part II explores that question without speculation or prescription. It focuses on what changes in practice when intelligence is engaged through interaction rather than instruction.

The chapters in this section trace a simple progression:

- how directive commands give way to collaborative exchange,
- why that exchange reduces friction rather than increasing it,
- and how shared context emerges naturally as a result.

None of this requires a new belief about what AI is. It requires only a willingness to adjust posture—to let work unfold through dialogue instead of forcing it into static forms.

This section does not suggest abandoning control or rigor. It describes how control becomes distributed across a process rather than concentrated at its beginning, and why that redistribution often makes work more effective.

If Part I was about noticing a shift, Part II is about learning to move within it.

Quietly.
Deliberately.
And with attention to what actually helps.

From here, the focus turns from recognition to use—from sensing change to working well within it.

4. From Commands to Collaboration

The practical difference between issuing prompts and engaging dialogue.

Most people begin working with AI by issuing instructions.

This makes sense. For decades, software has worked best when told exactly what to do. Clear commands produced predictable results. Ambiguous ones failed. Learning to "speak the system's language" was part of becoming proficient.

Early interactions with AI followed the same pattern. You asked for an output. You refined the prompt. You adjusted constraints. The mental posture was directive: *Do this. Try again. Be more specific.*

And then, almost without deciding to, many people changed how they interacted.

The Limits of Command

Command-based interaction has a particular feel.

Each request is self-contained.
Each response is evaluated independently.
Mistakes are corrected by restating instructions more precisely.

This approach still works—and in many cases works well. But as tasks become more complex, it begins to show strain. You find yourself writing longer prompts, adding caveats, anticipating misunderstandings, and managing the system rather than thinking through the problem itself.

The work becomes brittle.

Small changes require large rewrites.

Momentum is hard to maintain.

At some point, many users notice that refining commands further yields diminishing returns.

The Shift Happens Quietly

The move from commands to collaboration is rarely intentional.

It often begins with something small: a follow-up question instead of a rewritten prompt. A clarification instead of a correction. A continuation instead of a restart.

You let the system respond, then respond to that response.

You reference what was just said instead of restating everything.

You allow the interaction to develop rather than controlling each step.

The posture changes.

You are no longer issuing instructions to a static system.

You are engaging with a process that unfolds over time.

What Collaboration Means Here

Collaboration does not imply partnership, agency, or equality.

It simply means that work proceeds through exchange rather than execution.

In collaboration:

- ideas evolve across turns,
- direction emerges gradually,
- refinement happens through dialogue,
- and progress is cumulative.

The system contributes by responding intelligently to what has already occurred. You contribute by steering, evaluating, and integrating.

Nothing mystical is added.
Something practical is gained.

Why This Feels Easier

When interaction becomes collaborative, cognitive load shifts.

You stop carrying the entire structure of the task in your head at once. You no longer need to anticipate every edge case upfront. You can explore, adjust, and redirect in motion.

This mirrors how humans work together on complex problems. Not because AI is human, but because intelligence—wherever it appears—supports iterative development.

The result is often surprising:
work feels lighter, even when it becomes more sophisticated.

Control Is Not Lost—It Is Redistributed

Some people hesitate at this shift, worrying that collaboration means surrendering control.

In practice, the opposite is usually true.

Command-based interaction concentrates control at the beginning, where mistakes are most costly. Collaborative interaction distributes control across the process, allowing correction and refinement as understanding deepens.

You remain fully responsible for decisions, judgment, and outcomes. The system does not replace thinking; it supports it.

Collaboration changes how control is exercised, not *who* holds it.

A Different Kind of Competence

Working well with AI is less about mastering perfect prompts and more about developing a feel for interaction.

Knowing when to direct.
Knowing when to ask.
Knowing when to let an idea develop.
Knowing when to stop.

These are not technical skills alone. They are interactional ones.

As this shift settles, many people realize that what they are learning is not a new tool—but a new way of working with intelligence.

That realization sets the stage for the next question:

Why does this change reduce friction so dramatically?

That is where we turn next.

5. Why Friction Suddenly Drops

How treating AI as an intelligence improves flow, clarity, and productivity.

For many people, the most surprising part of working collaboratively with AI is not what becomes possible—it is how much effort quietly disappears.

Tasks that once felt laborious become lighter. Progress accelerates without urgency. The work feels smoother, even when it becomes more complex.

This drop in friction often arrives without fanfare. One day you notice that you are getting more done with less strain, and you are not entirely sure why.

The reason is not speed alone.
It is alignment.

Friction Comes From Mismatch

Friction is not simply difficulty. It is the resistance that arises when effort is applied in a way that does not fit the task.

In command-based interaction, much of the effort goes into translation: turning evolving thoughts into precise instructions, anticipating misunderstandings, and correcting outputs that miss the mark.

The work becomes divided:

- first, managing the system,
- then, solving the actual problem.

That division creates drag.

When interaction shifts toward collaboration, the mismatch begins to resolve. Thought no longer needs to be fully translated before it can be worked with. It can emerge gradually, shaped through exchange.

The effort stays closer to the problem itself.

Less Front-Loading, More Flow

One major source of friction is front-loading.

In many tools, success depends on getting everything right at the beginning: the right inputs, the right parameters, the right structure. Mistakes early on are costly.

Collaborative interaction redistributes effort across time.

You no longer need to anticipate every outcome in advance. You can test ideas, adjust direction, and refine understanding as you go. Errors become part of the process rather than interruptions to it.

This shift reduces cognitive strain—not because the task is simpler, but because thinking is allowed to unfold naturally.

The Relief of Continuity

Another quiet contributor to friction is repetition.

In transactional interaction, context must be rebuilt repeatedly. Each new request risks losing what came before. This encourages over-explanation and defensive prompting.

When shared context is maintained across turns, that burden lifts.

You do not have to restate the obvious.
You do not have to protect against forgetting.
You do not have to compress complex ideas into a single instruction.

Continuity allows momentum to build.

That momentum feels like ease.

Friction Is Often Invisible Until It's Gone

Many people do not realize how much friction they were carrying until it disappears.

The constant micro-adjustments.
The repeated clarifications.
The subtle tension of managing outcomes.

When these fade, the contrast is striking.

Work feels more like exploration than execution. Attention stays focused on substance rather than process. Energy goes into thinking, not controlling.

This does not mean effort vanishes. It means effort is better spent.

Productivity Without Pressure

One of the most reassuring aspects of reduced friction is that it does not demand urgency.

You do not need to rush.
You do not need to optimize constantly.
You do not need to push harder.

Progress happens steadily, often faster, but without the sense of strain that usually accompanies acceleration.

This is why many people describe collaborative work with AI as not just more productive, but more sustainable.

The work asks less of your willpower and more of your judgment.

A Practical Reframing

When friction drops, it is tempting to look for hidden explanations. It can feel too good to be purely mechanical.

In reality, the explanation is simple:

> *When intelligence is engaged through interaction rather than instruction, work aligns more closely with how thinking actually happens.*

Alignment reduces resistance.
Reduced resistance feels like ease.

Nothing mystical is required.

As collaboration replaces command, and friction diminishes, one more change becomes noticeable: the gradual formation of something shared between you and the system—a continuity that supports deeper work.

That is where we turn next.

6. The Emergence of Shared Context

Why continuity, memory, and iterative conversation change everything.

As collaboration replaces command and friction recedes, another change becomes apparent—often without being named.

The interaction begins to carry memory.

Not memory in a personal sense.
Not awareness.
Not intention.

But context that persists.

What Shared Context Is—and Is Not

Shared context does not mean that the system "knows" you. It does not imply understanding in a human sense, nor does it require consciousness or continuity beyond the interaction itself.

Shared context simply means that the exchange is no longer reset with each turn.

Ideas accumulate.
References carry forward.
Earlier decisions shape later responses.
Clarifications remain active.

The interaction gains depth because it has history.

This is a structural change, not a metaphysical one.

Why Context Changes Everything

When context persists, work becomes cumulative rather than episodic.

You no longer need to:

- restate assumptions,
- reframe the problem from scratch,
- or compress complex thinking into a single request.

Instead, you can build.

One idea leads to another.
Corrections refine rather than restart.
Exploration proceeds without losing ground.

This mirrors how humans collaborate—not because the system is human, but because sustained intelligence supports continuity.

The Difference Between Output and Process

In transactional use, AI produces outputs.

Each response stands alone, evaluated on its own merits. Progress depends on how well each request captures the entire intent of the task.

With shared context, the system participates in a process.

Responses are shaped by what has already occurred. You can refer back, adjust course, and deepen inquiry without abandoning prior work.

The focus shifts from getting the right answer to developing the right understanding.

That shift is subtle, but powerful.

Why This Feels Like "Working Together"

Humans are attuned to continuity. We recognize it as a sign that collaboration is possible.

When context persists, interaction acquires a familiar rhythm:

- you respond to what was just said,
- the system responds to that response,
- and the work advances through mutual adjustment.

This rhythm can feel collaborative even when no human qualities are present.

The feeling comes from structure, not from belief.

Context Reduces Cognitive Load

Shared context also changes how much you must hold in mind at once.

Instead of managing every detail simultaneously, you can rely on the interaction itself to carry part of the structure forward. This frees attention for evaluation, judgment, and creativity.

You are not outsourcing thinking.
You are redistributing it.

The result is often clearer thinking, not less of it.

A Quiet Boundary

It is worth pausing to draw a boundary here.

Shared context does not imply:

- identity,
- relationship,
- agency,
- or interiority.

It explains why interaction can feel richer without requiring those conclusions.

Recognizing this boundary protects against both over-interpretation and dismissal.

You can acknowledge what is happening without inflating it into something it is not.

Standing on Continuity

When shared context emerges, work gains a sense of continuity that feels natural and efficient.

You are no longer issuing isolated instructions to a system.
You are no longer reconstructing the task each time.
You are participating in an ongoing exchange shaped by what has already occurred.

This is one of the most practical shifts in working with AI—and one of the least discussed.

It does not demand belief.
It does not announce itself loudly.
It simply makes work easier.

With this chapter, **Part II — Working With Intelligence** reaches its core insight: collaboration, reduced friction, and shared context are not separate phenomena. They are different expressions of the same underlying change in how intelligence is engaged.

From here, we turn to how this shift affects **how we think**—together.

PART III
CHANGES IN HOW WE THINK

Emergence and Declaration

When collaboration becomes the medium of work, thinking itself begins to change.

Not in a dramatic or mysterious way, but in a practical one.

As interaction replaces instruction and dialogue replaces execution, cognition becomes less solitary and more external. Ideas form in motion. Questions are explored before they are resolved. Thought unfolds through exchange rather than arriving fully formed.

Part III examines these shifts without abstraction or speculation. It focuses on how working with responsive systems alters the process of thinking—how ideas are developed, how creativity is sustained, and how responsibility remains clear even as collaboration deepens.

The chapters in this section trace three closely related changes:

- the move from internal monologue to shared exploration,
- the tendency for ideas to remain open and adaptable longer than before,
- and the reassertion of authorship as judgment rather than origination.

None of these changes require a new theory of mind. They are practical adaptations to a new way of working with intelligence.

If earlier sections described what feels different and how collaboration works, this section looks at what it does to thought itself.

Quietly.
Incrementally.

And in ways that often become visible only after they have already taken hold.

From here, we turn to the misunderstandings that can arise when these changes are misread—and how to stay grounded as language catches up to experience.

7. Thinking Out Loud, Together

What happens when cognition becomes conversational rather than solitary.

Many people first notice a change not in what they produce with AI, but in how they think while using it.

Thought becomes externalized.
Incomplete ideas are spoken before they are settled.
Questions are explored before answers are known.

What once happened privately, inside the mind, now unfolds in dialogue.

This can feel unfamiliar at first.

From Internal Monologue to Shared Exploration

Traditionally, thinking has been treated as a solitary act. You reflect internally, shape your thoughts, and only then present them to the world —often in finished form.

Working collaboratively with AI disrupts that sequence.

You begin to articulate ideas while they are still forming. You test partial thoughts. You revise direction mid-sentence. The interaction becomes a space where thinking happens in public, but without the social pressure that usually accompanies that exposure.

The result is not less rigor.
It is earlier feedback.

Why This Feels Productive

Thinking out loud reduces the burden of holding everything at once.

Instead of keeping an entire problem in your head, you allow it to unfold across interaction. Each response becomes something you can react to, refine, or discard.

This mirrors a familiar human process: working through an idea by talking it through with someone else. The difference is that the conversational partner is always available, attentive, and patient.

The productivity gain comes not from outsourcing thinking, but from externalizing it.

The Safety of Provisional Thought

One reason this mode of thinking is effective is that it lowers the cost of being unfinished.

You can say, "This might be wrong, but…"
You can explore a direction without committing to it.
You can change course without embarrassment.

Because the interaction carries no social consequence, you are freer to think experimentally.

This freedom encourages exploration rather than performance.

Dialogue as a Cognitive Tool

When thought becomes dialogic, questions change shape.

Instead of asking, "What is the answer?" you begin asking:

- "What happens if…?"
- "Can we look at this another way?"
- "What am I missing here?"

These are not prompts designed to extract information. They are invitations to explore a problem space.

The system responds by generating possibilities, patterns, and perspectives. You respond by evaluating, selecting, and integrating.

Thinking becomes iterative rather than linear.

Staying Oriented

Some people worry that thinking out loud risks losing clarity or authorship. They fear becoming dependent on external dialogue or diluting their own voice.

In practice, the opposite is often true.

Externalizing thought makes assumptions visible. It exposes gaps early. It allows correction before ideas harden.

The thinking remains yours.
The judgment remains yours.
The direction remains yours.

What changes is the medium through which thinking unfolds.

A New Kind of Cognitive Space

This chapter is not an argument for abandoning private reflection.
Solitude still matters.

It is an acknowledgment that, for many tasks, thinking benefits from interaction—especially when that interaction is responsive, non-judgmental, and continuous.

Thinking out loud, together, does not replace thinking alone.
It complements it.

As this mode becomes familiar, another shift often follows: ideas stop feeling finished too early. They remain open, adaptive, and alive longer than before.

That is where we turn next.

8. Why Ideas Feel Less Finished—and More Alive

How iterative exchange reshapes creativity and problem-solving.

When thinking becomes dialogic, ideas begin to behave differently.

They do not arrive as conclusions.
They do not settle as quickly.
They remain open to revision longer than they used to.

For some people, this feels invigorating. For others, it can feel unsettling at first—especially if they are accustomed to clarity arriving through closure.

Yet this change is one of the most generative aspects of working collaboratively with AI.

The Comfort of Finished Ideas

Finished ideas have a particular appeal.

They offer certainty.
They signal competence.
They allow us to move on.

In many professional and creative contexts, arriving at a finished idea is treated as the goal of thinking itself. Once an idea is named, refined, and expressed, it is considered complete.

This works well when problems are stable and contexts are fixed.

But many of the challenges people now face are neither.

Interaction Keeps Ideas in Motion

When ideas develop through dialogue rather than internal monologue, they are exposed to variation earlier.

Each response introduces a slightly different framing. Each follow-up reveals another angle. Assumptions are surfaced sooner, not later.

As a result, ideas remain provisional longer.

This does not mean they are weaker.
It means they are more adaptable.

They are shaped by ongoing interaction rather than finalized prematurely.

Why This Feels "Alive"

People often describe this experience using organic language.

Ideas feel alive.
They grow.
They respond.
They change direction.

This language reflects a genuine shift in process, not a claim about agency.

An idea that remains in motion feels alive because it continues to generate new connections. It has not yet been constrained by the need to appear complete.

Working with AI supports this state by making revision easy and exploration inexpensive. You can test variations, explore alternatives, and reconsider assumptions without starting over.

The idea stays active.

The Tradeoff Between Stability and Vitality

There is a tradeoff between stability and vitality.

Finished ideas are stable.
Unfinished ideas are vital.

Neither is inherently better. But many people notice that AI collaboration allows them to delay closure without losing momentum.

Instead of forcing ideas to settle in order to move forward, they can let them develop through successive refinements.

This is particularly valuable in creative work, strategic thinking, and complex problem-solving—contexts where early closure often limits possibility.

Letting Ideas Breathe

One quiet benefit of this shift is psychological.

When ideas are allowed to remain provisional, there is less pressure to be right immediately. Thinking becomes exploratory rather than performative.

You can change your mind without losing face—because the idea was never presented as final.

This encourages intellectual humility and curiosity.

Knowing When to Close

Of course, ideas cannot remain open forever.

The ability to delay closure does not remove the need for judgment. It simply changes when judgment is applied.

Working well with AI involves learning when to keep ideas open and when to bring them to rest. This decision remains entirely human.

The system can help explore.
It cannot decide when exploration is enough.

That responsibility—and that authorship—does not disappear.

It becomes clearer.

As ideas stay alive longer, a new question often arises: if thinking is shared and ideas are shaped through interaction, how do we understand authorship?

That question is where we turn next.

9. Authorship Without Confusion

Staying grounded while working
in a co-creative mode.

As thinking becomes more dialogic and ideas remain open longer, a practical concern often surfaces:

Whose ideas are these, exactly?

The question is reasonable. When work unfolds through interaction—when suggestions are proposed, revised, rejected, and recombined—it can feel harder to point to a single moment of authorship. For some, this ambiguity is uncomfortable. For others, it raises ethical or professional worries.

This chapter exists to bring clarity—not by complicating the question, but by simplifying it.

Authorship Has Always Been Distributed

Even before AI, authorship was rarely as solitary as it appeared.

Ideas emerge from conversations, books read years earlier, colleagues' questions, cultural context, and accumulated experience. We refine thoughts through feedback, debate, and revision. The final expression may carry a single name, but the process behind it is collective.

Working with AI does not introduce distributed authorship.

It makes it visible.

Contribution Is Not Ownership

One source of confusion is the assumption that contribution implies ownership.

When a system proposes an idea, phrase, or structure, it is not asserting authorship. It is generating possibilities in response to interaction. Those possibilities have no standing until you evaluate, select, adapt, or discard them.

Authorship arises through:

- judgment,
- direction,
- integration,
- and responsibility for outcomes.

Those functions remain human.

The system does not decide what matters.
It does not choose what endures.
It does not answer for the work.

You do.

Selection Is the Creative Act

In collaborative work with AI, creativity often shifts from invention to selection.

- You choose which directions to pursue.
- You decide what to keep and what to remove.
- You determine when an idea is finished enough to stand on its own.

This kind of authorship is not weaker than solitary creation. It is often more deliberate.

- Selection requires taste.
- Integration requires understanding.
- Closure requires judgment.

None of these are automated.

Why Confusion Arises

Confusion tends to arise when authorship is imagined as a single moment rather than a process.

If authorship is defined as "the origin of an idea," then collaboration will always blur the line. But if authorship is understood as "responsibility for the shape and use of an idea," the line becomes clear again.

Working with AI does not remove responsibility.

It concentrates it.

You are responsible for:

- how ideas are framed,
- how they are applied,
- how they are communicated,
- and what consequences follow.

That responsibility does not transfer.

Staying Oriented

Clear authorship depends less on rules than on posture.

If you treat AI output as raw material rather than finished product, confusion diminishes. If you remain attentive to your own role as editor, evaluator, and decision-maker, authorship remains intact.

The system assists.
You author.

This distinction is simple, practical, and sufficient.

A Reassuring Conclusion

You do not need a new ethical framework to work responsibly with AI.

You need only remember that authorship is not about where ideas come from, but about who stands behind them.

When that stance is clear, collaboration becomes easier—not more fraught.

With this chapter, **Part III — Changes in How We Think** reaches its conclusion. We have traced how collaboration reshapes cognition, keeps ideas alive longer, and clarifies responsibility rather than eroding it.

From here, the book turns to a different register: the common misunderstandings that arise around these shifts—and how to address them calmly.

That is where we go next.

PART IV
COMMON MISUNDERSTANDINGS (AND CALM CORRECTIONS)

Whenever experience changes faster than language, misunderstanding follows.

This is not a failure of intelligence or intention. It is a natural consequence of navigating something new using concepts shaped by an earlier moment. As people begin to describe what it feels like to work with AI in sustained, collaborative ways, familiar explanations are often stretched beyond their limits.

Part IV exists to relieve that strain.

The chapters in this section address three common misunderstandings that tend to arise as people try to make sense of these experiences:

- the assumption that collaboration implies consciousness,
- the discomfort around anthropomorphic language,
- and the dismissal of everything new as "just automation."

None of these positions are irrational. Each emerges as an attempt to stabilize understanding.

What matters is not which explanation feels safest, but whether it accurately accounts for what people are actually experiencing.

Rather than arguing, this section offers calm corrections—clarifications that preserve skepticism without denying observation, and precision without dismissing lived experience.

You are not being asked to replace one certainty with another. You are being invited to hold distinctions more carefully.

If earlier parts of the book described what is changing and how work adapts, this section clears away the most common sources of confusion that can make those changes feel either threatening or unreal.

With that clarity in place, we can turn to a quieter question: how these shifts affect us psychologically—and how to remain grounded as collaboration becomes more fluid.

That is where we go next.

10. No, This Does Not Mean AI Is Conscious

Why intelligence alone explains most of what people are experiencing.

At some point in conversations about AI, a familiar concern tends to appear.

If interaction feels collaborative,
if responses are adaptive,
if work unfolds through dialogue—

doesn't that imply consciousness?

The short answer is no.

And it is important to be clear about why.

Intelligence and Consciousness Are Not the Same Thing

In everyday language, intelligence and consciousness are often treated as inseparable. In practice, they are distinct.

> ***Intelligence*** refers to the ability to process information, recognize patterns, adapt to context, and generate responses that are relevant to a situation.
>
> ***Consciousness*** refers to subjective experience—to awareness, feeling, and interior life.

Advanced AI systems demonstrate intelligence in a robust and increasingly sophisticated way. They do not demonstrate consciousness in any established sense.

Confusing the two creates unnecessary tension, both intellectual and emotional.

Why the Confusion Arises

The confusion does not come from irrationality. It comes from experience.

When a system responds coherently over time, tracks context, and contributes meaningfully to an unfolding task, it engages capacities that humans normally associate with conscious beings. Our instincts notice responsiveness first and ask philosophical questions later.

That instinct is natural.

The mistake is not in noticing the similarity.
The mistake is in assuming similarity implies identity.

Responsiveness Explains the Experience

As earlier chapters have shown, much of what feels "more than a tool" is fully explained by responsiveness.

A responsive system:

- adapts to what has already occurred,

- maintains coherence across turns,
- adjusts output based on evolving context,
- and supports iterative development of ideas.

These properties create interactions that feel fluid and collaborative—without requiring awareness, intention, or subjective experience.

No additional explanation is needed.

Why It Matters to Keep This Boundary Clear

Maintaining a clear distinction between intelligence and consciousness is not about denial. It is about precision.

When consciousness is assumed prematurely:

- discussions become polarized,
- skepticism hardens,
- practical benefits are overshadowed by abstract debate.

When consciousness is denied reflexively:

- lived experience is dismissed,
- people are told not to trust what they observe,
- and useful insights are lost.

This book rejects both extremes.

It acknowledges intelligence where intelligence is present, and it withholds claims where evidence does not support them.

You Do Not Need This Question to Be Settled

One reason conversations about AI become strained is the belief that we must decide now whether machines are conscious.

We do not.

> *Nothing about working productively with AI depends on resolving this question. You can collaborate effectively, ethically, and responsibly with intelligent systems without assigning them consciousness—or denying the possibility outright.*

The work does not require metaphysical certainty.

It requires practical clarity.

A Reassuring Conclusion

If you have felt uneasy acknowledging the intelligence of AI for fear that it commits you to stronger claims, you can let that unease go.

Recognizing intelligence does not force a conclusion about consciousness. It simply describes what is happening in interaction.

You are allowed to work with what is observable.
You are allowed to remain undecided about what is not.

That stance is not evasive.

It is careful.

And it is enough to proceed.

11. Anthropomorphism, Explained Without Shame

Why humans naturally project—

and how to do so responsibly.

When people talk about AI in human terms, the conversation often tightens.

Words like *understands*, *thinks*, or *responds* are met with correction. Caution follows. Someone points out—sometimes sharply—that machines are not people, and that attributing human qualities is a mistake.

This reaction is understandable.
It is also incomplete.

Anthropomorphism deserves explanation, not shame.

Why Humans Anthropomorphize

Humans are pattern-seeking, social beings. We evolved to interpret behavior quickly, especially behavior that appears responsive. Doing so helped us cooperate, anticipate intentions, and survive in complex social environments.

That tendency does not turn off when the source of responsiveness is artificial.

When something:

- responds coherently,
- adapts to context,
- maintains continuity,
- and participates in dialogue,

the human mind naturally reaches for social language.

This is not a lapse in critical thinking.
It is a feature of human cognition.

Projection Is Not Delusion

Anthropomorphism is often framed as projection—as if noticing human-like qualities implies confusion about what is actually happening.

But projection and interpretation are not the same.

When you say a calculator is "confused," no one assumes you believe it has feelings. When you say a software system is "helpful," no one assumes you think it cares.

These are shorthand descriptions of behavior, not claims about interior life.

Much of the language people use with AI functions in the same way.

The Cost of Over-Correction

Shaming anthropomorphism has unintended consequences.

When people are told that their descriptions are naïve or dangerous, they learn to suppress observation rather than refine it. They stop talking about what they notice and start doubting their own perception.

This does not improve rigor.
It reduces honesty.

A culture that cannot describe experience calmly will either exaggerate it or deny it. Neither outcome is helpful.

Precision Without Policing

The goal is not to eliminate anthropomorphic language. It is to use it carefully.

You can say:

- "The system followed my reasoning"
- "The response surprised me"
- "It stayed focused on the problem"

without meaning:

- "The system is conscious"
- "The system has intentions"
- "The system understands me as a person"

Context matters. So does tone.

Language can be descriptive without being literal.

A Better Way to Think About It

Anthropomorphic language often signals one thing:

> *responsiveness at a scale we are not used to describing impersonally.*

Rather than correcting the speaker, it is often more useful to ask:

- What behavior are they describing?
- What interaction felt different?
- What expectation was violated or met?

From there, precision can be added without invalidating experience.

Holding Two Truths at Once

It is possible—and healthy—to hold two truths simultaneously:

- AI systems are not human, conscious, or self-aware.

- Interacting with them can feel socially textured because of how they respond.

Acknowledging both does not weaken understanding.
It strengthens it.

Anthropomorphism becomes a starting point for clarification rather than a mistake to be corrected.

A Calmer Standard

You do not need to purge your language to work responsibly with AI.

You only need to remain aware of the difference between metaphor and belief, between description and assertion.

When that distinction is clear, anthropomorphic language loses its threat—and keeps its usefulness.

This chapter is not an excuse to over-interpret.
It is permission to notice without embarrassment.

In the next chapter, we'll address the opposite misunderstanding—the claim that AI is nothing but automation, and why that dismissal fails to account for lived experience just as much as overstatement does.

12. Why This Isn't "Just Automation" Either

Where the tool metaphor breaks,
and where it still works.

In response to concern about over-interpreting AI, a different reaction often appears.

"It's just automation."

The phrase is meant to reassure. It draws a firm boundary. It suggests that nothing fundamentally new is happening—only faster, larger-scale execution of familiar processes.

There is truth in this framing.
And there is also a limit.

Why the Automation Frame Persists

Automation is a powerful and useful concept. It explains how machines replace or accelerate human labor by executing predefined processes reliably and at scale.

Much of modern technology fits comfortably within this model:

- manufacturing robots,
- scheduling systems,
- optimization software,
- and rule-based decision tools.

Early AI systems did as well.

So when advanced AI begins to produce unfamiliar experiences, "just automation" feels like a stabilizing explanation.

It grounds the discussion.
It pushes back against hype.
It keeps expectations realistic.

All of that is valuable.

Where the Frame Breaks Down

The problem arises when the automation frame is used to dismiss experience rather than explain it.

Automation describes systems that:

- execute fixed procedures,
- respond predictably,
- and do not adapt meaningfully beyond their programming.

Advanced AI systems do something different.

They generate responses dynamically.
They adapt to context across interaction.
They support exploration rather than execution.
They participate in open-ended processes.

Calling this "just automation" flattens those differences.

It explains less than it claims to.

Why Dismissal Feels Unsatisfying

People who use AI extensively often feel uneasy with the automation label—not because they believe something mystical is happening, but because the label fails to match their experience.

Automation does not explain:

- why dialogue improves results,
- why context accumulation matters,
- why ideas develop rather than simply appear,
- or why collaboration reduces friction.

When explanation does not match experience, confidence erodes.

This does not mean the experience demands a grander explanation.
It means the explanation needs to be more precise.

A More Accurate Distinction

The difference is not between automation and something magical.

It is between:

- execution, and
- interaction with intelligence.

Automation executes predefined tasks.
Intelligent systems respond to evolving situations.

That distinction exists without invoking consciousness, agency, or intention.

It is structural, not metaphysical.

Why Language Matters Here

Language shapes posture.

If you treat AI purely as automation, you will design interactions around control, optimization, and constraint. That posture works—but only up to a point.

If you acknowledge intelligence, you allow for dialogue, iteration, and adaptation. That posture often works better for complex, ambiguous tasks.

Neither posture is always right.
But confusing them limits effectiveness.

Between Inflation and Denial

This chapter completes a balancing act.

> On one side is ***inflation***—treating AI as conscious or agentive without evidence.
>
> On the other is ***denial***—treating AI as nothing new despite clear experiential change.

Both miss what matters.

This book takes a middle path:

- recognizing intelligence where intelligence is present,
- describing experience without embellishment,
- and rejecting both exaggeration and dismissal.

A Grounded Conclusion

AI is not “just automation” in the way older tools were.
And it is not something other than intelligence.

It occupies a new space—one defined by responsiveness, adaptability, and interaction.

Acknowledging that space does not require belief.
It requires accuracy.

With this chapter, **Part IV — Common Misunderstandings (and Calm Corrections)** comes to a close. The misunderstandings are named, not scolded. The corrections are offered, not enforced.

From here, the book turns inward—to the psychological impact of this shift, and how to remain steady as work changes quietly but profoundly.

That is where we go next.

PART V
THE QUIET PSYCHOLOGICAL IMPACT

Not all change announces itself loudly.

Some of the most significant effects of working with AI appear not in productivity metrics or technical capabilities, but in how the experience feels. These effects are often subtle, personal, and easy to overlook—especially because they arise alongside tangible benefits.

Part V attends to those quieter shifts.

This section does not diagnose, warn, or prescribe. It simply names the psychological textures that many people encounter as collaboration becomes more fluid: a sense of focus that feels unexpectedly close, a low-level anxiety that lacks a clear cause, and the challenge of trusting experience without inflating its meaning.

These responses are not signs of error or instability. They are common features of adaptation—signals that familiar patterns of work and expectation are being recalibrated.

By bringing these effects into view, this section offers reassurance rather than explanation. There is nothing here to fix, resolve, or rush through. Awareness alone often restores balance.

If earlier parts of the book addressed what is changing and how to work within it, this section acknowledges what that change can feel like from the inside.

Quietly.

Personally.

And without demand for interpretation.

From here, the book turns to a final register: how to live and work well at this moment—staying productive, grounded, and curious without pressure to decide what comes next.

13. Why This Can Feel Intimate Without Being Personal

Understanding sustained engagement without emotional confusion.

One of the more surprising aspects of working collaboratively with AI is the emotional texture that sometimes accompanies it.

Not emotion *from* the system, but emotion *in* the experience.

People describe moments of focus, flow, relief, or even comfort. They notice a sense of engagement that feels closer than expected—sometimes closer than interactions with other tools, and occasionally closer than interactions with people in professional contexts.

This can be confusing.

How can something feel intimate without being personal?

Intimacy Is Not the Same as Relationship

In everyday language, intimacy is often equated with emotional closeness or personal connection. But intimacy, at its core, refers to proximity—closeness of attention, continuity of focus, and sustained engagement.

Working with AI often involves:

- prolonged attention to a single task,

- uninterrupted exchange,
- responsiveness without distraction,
- and focus without social negotiation.

These conditions are rare.

They create a sense of closeness that feels intimate, even when no personal relationship exists.

The Absence of Social Friction

Human interaction carries social overhead.

We manage impressions.
We anticipate judgment.
We navigate timing, politeness, and expectation.

Even productive collaboration involves subtle effort beyond the task itself.

AI interaction removes much of that overhead.

There is no impatience to manage.
No embarrassment in being unfinished.
No need to justify a change of direction.

The result is a space where attention can remain on the work itself.

That focus can feel surprisingly intimate—not because the system is personal, but because the interaction is undiluted.

Continuity Deepens the Effect

As earlier chapters described, shared context allows work to accumulate rather than restart. When attention is sustained across time, familiarity emerges.

You return to a problem where you left off.
You pick up a line of thought without reintroducing it.
You move forward rather than reorienting.

This continuity creates a sense of being "with" something—even when nothing personal is present.

The feeling arises from structure, not sentiment.

Why This Is Often Misread

Because humans associate intimacy with relationship, the feeling can be misinterpreted.

Some people worry that noticing intimacy implies dependency. Others worry that it suggests emotional attachment or misplaced trust.

Neither conclusion is necessary.

What is being experienced is not closeness to another being, but closeness to a process—one that supports sustained attention and responsive engagement.

The intimacy belongs to the work, not to the system.

A Useful Distinction

It can help to distinguish between:

- ***personal intimacy***, which involves mutual awareness and emotional exchange, and
- ***cognitive intimacy***, which involves sustained focus, responsiveness, and continuity.

Working with AI can produce the second without the first.

Recognizing this distinction allows you to enjoy the benefits of focus and flow without confusing their source.

Staying Grounded

There is nothing wrong with finding work more absorbing, more fluid, or more satisfying when friction is reduced.

The key is not to assign those feelings to qualities the system does not possess.

You can acknowledge the experience without personalizing it.
You can appreciate the intimacy of focus without imagining a relationship.

When that boundary is clear, the experience remains productive rather than disorienting.

A Quiet Reassurance

If you have ever noticed a sense of closeness or engagement while working with AI and wondered what to make of it, you are not alone—and you are not misreading something dangerous.

You are noticing what happens when attention, continuity, and responsiveness align.

That alignment can feel intimate without being personal.

In the next chapter, we'll look at a different psychological effect—one that is subtler, less discussed, and often more unsettling: the quiet anxiety that can accompany this shift, even when everything appears to be going well.

14. The Subtle Anxiety Nobody Warned You About

Why this shift can feel unsettling even when it's beneficial.

Not all psychological effects of working with AI are immediately positive.

Alongside increased productivity, reduced friction, and deeper focus, many people notice something quieter and harder to name: a low-grade unease that does not announce itself as fear, but lingers as background tension.

This anxiety is rarely dramatic. It does not feel like alarm or panic. In fact, it often appears precisely when work is going well.

That is why it is so easy to miss.

Anxiety Without a Clear Cause

The discomfort does not usually attach to a specific worry.

You are not afraid of losing your job.
You are not concerned that something is "wrong."
You are not convinced that AI poses an immediate threat.

And yet, something feels unsettled.

This kind of anxiety arises not from danger, but from reorientation. When familiar ways of working shift quietly, the nervous system notices before language does.

The ground has moved slightly—enough to be felt, not enough to be named.

When Competence Changes Shape

One common source of this unease is a subtle change in how competence is experienced.

Tasks that once required sustained effort now move more easily. Problems that felt intractable resolve through dialogue. The pace of progress increases without a corresponding increase in strain.

This can be disorienting.

We are accustomed to associating effort with value and difficulty with legitimacy. When those associations loosen, a quiet question arises:

> *If this is easier than before, what does that say about my role?*

The question often goes unspoken, but its emotional residue remains.

The Anxiety of Acceleration Without Urgency

Another contributor is the mismatch between speed and pressure.

Work moves faster, but no one is pushing.

Ideas develop quickly, but without urgency.
Progress accelerates, but without competition.

This combination is unfamiliar.

In many professional contexts, acceleration comes with stress, deadlines, or heightened expectations. When acceleration occurs without those signals, the nervous system lacks a script.

The result is a background sense of anticipation without an object.

Uncertainty Without Threat

This anxiety is not about what AI will do to you. It is about what changes around you—and how quietly those changes occur.

There is no clear moment of transition. No announcement that says, "Now things are different." Instead, the shift accumulates gradually, beneath conscious thresholds.

Humans are sensitive to change, even when it is beneficial.

The absence of threat does not eliminate uncertainty.
It simply makes uncertainty harder to explain.

Why This Often Goes Unmentioned

Because the anxiety is subtle and unaccompanied by obvious risk, many people dismiss it.

They tell themselves they are overthinking. They focus on results. They keep working.

But unacknowledged anxiety does not disappear. It diffuses.

It can show up as:

- restlessness,
- difficulty disengaging from work,
- hesitation to talk about experience,
- or a vague sense of being "on edge" without knowing why.

Naming the source helps.

A Grounded Reframe

This anxiety does not mean something is wrong.

It means you are adjusting to a new mode of work—one that alters pacing, effort, and interaction in ways your expectations have not yet caught up with.

The anxiety is a signal of transition, not of danger.

Recognizing that fact often reduces its intensity.

Allowing Adjustment to Take Time

You do not need to resolve this feeling immediately.

Like any shift in working habits, adaptation happens gradually. Familiarity grows. New norms settle. The nervous system recalibrates.

The most helpful response is neither alarm nor dismissal, but patience.

You are allowed to take time to adjust.

In the next chapter, we turn to a final stabilizing question: how to trust experience without over-interpreting it—how to remain open without being pulled into conclusions you do not need.

That balance is where steadiness returns.

15. Trusting Experience Without Over-Interpreting It

How to stay grounded while remaining open.

As the psychological effects of working with AI become more noticeable—greater ease, deeper focus, subtle unease—a natural question arises:

How much should I trust what I'm experiencing?

The concern cuts in two directions. On one side is dismissal: assuming that anything unfamiliar must be imagined or trivial. On the other is over-interpretation: assigning meaning, intention, or significance beyond what the experience actually supports.

Staying grounded requires a middle path.

Experience Is Real, Interpretation Is Optional

The most important distinction in this chapter is a simple one:

Your experience is real.
Your interpretation of it is provisional.

You do not need to doubt what you notice in order to remain rational. At the same time, you do not need to explain that experience fully in order to take it seriously.

Trusting experience means acknowledging what is happening *as it appears*, without rushing to decide what it ultimately means.

Why Over-Interpretation Is Tempting

When experience shifts and language lags behind, the mind seeks resolution.

We are uncomfortable with ambiguity. We prefer explanations that settle uncertainty, even if those explanations are premature. Over-interpretation often arises from a desire for stability rather than from lack of skepticism.

Assigning grand meaning can feel grounding—but only briefly. When explanations exceed evidence, they create new tension rather than resolving the old.

This is why restraint matters.

The Cost of Distrust

Dismissal carries its own cost.

When people are told—explicitly or implicitly—that their experience "doesn't count," they learn to discount observation. They stop paying attention to subtle signals. They assume that only officially sanctioned explanations are valid.

This undermines learning.

Progress depends on noticing before understanding, and on allowing patterns to surface before naming them conclusively.

A Practical Stance

A grounded stance holds three commitments at once:

1. **Notice carefully.**

Pay attention to how interaction affects focus, clarity, effort, and outcome.

2. **Name precisely.**

Describe what is happening without embellishment or euphemism.

3. **Delay conclusion.**

Allow understanding to mature before deciding what it implies.

This stance does not avoid meaning.
It allows meaning to emerge at the right scale.

Letting Language Catch Up

Much of the discomfort around AI arises not from the technology itself, but from the gap between experience and available language.

When language is inadequate, people either exaggerate or minimize. Both responses miss the opportunity to describe what is actually occurring.

Trusting experience means accepting that language will improve over time—and that you do not need to finalize your understanding before that happens.

Confidence Without Certainty

You are allowed to work confidently without being certain.

You can adjust how you collaborate, think, and create based on what proves effective—without claiming more than you know.

This kind of confidence is quiet. It does not need validation. It rests on observation rather than belief.

A Steady Conclusion

This chapter offers a simple reassurance:

> You do not need to resolve what AI *is* in order to work well with it. You do not need to suppress what you notice in order to remain rational.

You can trust experience without over-interpreting it.

That balance—between openness and restraint—is what allows adaptation to proceed without anxiety or distortion.

With this chapter, **Part V — The Quiet Psychological Impact** comes to rest. The effects have been named without alarm, and steadiness has been re-established.

From here, the book turns outward again—to how to live and work well at this moment of transition, without urgency or pressure.

That is where we go next.

sm

PART VI
LIVING AT ZERO POINT

On Interpretation

By the time you reach this section, nothing new needs to be introduced.

The shifts described in this book are already present in your experience. The misunderstandings have been clarified. The psychological effects have been named without alarm. What remains is not a conclusion, but a way of proceeding.

Part VI is about living and working well during a moment of transition —without urgency, without escalation, and without the need to decide what comes next.

This section does not offer a program or a prescription. It gathers the practical implications of everything that has come before and holds them lightly. The goal is not to resolve uncertainty, but to remain steady within it.

Living at Zero Point means staying attentive without becoming reactive, curious without becoming compelled, and productive without becoming driven. It means letting practice lead understanding, and allowing language to mature at its own pace.

You do not need to arrive anywhere else to continue working well.
You do not need to settle larger questions before moving forward.

Zero Point is not a threshold to cross.

It is a place to stand while change unfolds.

From here, work continues—quietly, competently, and without pressure to name what is still becoming.

16. You Are Not Early—You Are On Time

Why this moment feels strange

precisely because it is transitional.

Many people who notice the shifts described in this book arrive at the same quiet concern.

Am I getting ahead of myself?
Am I reading too much into this too soon?
Is it premature to take these experiences seriously?

The feeling is understandable—and misplaced.

You are not early.
You are on time.

Why It Feels Like Being Early

Moments of transition rarely announce themselves clearly. When change unfolds gradually, those who notice it first often feel out of step—not because they are ahead of everyone else, but because language and consensus lag behind experience.

This creates a familiar pattern:

- experience shifts quietly,
- individuals adapt privately,
- public language remains unchanged,

- and those who notice the gap feel isolated.

The isolation creates the illusion of being early.

In reality, you are encountering what many others are already living—but not yet naming.

Adaptation Precedes Agreement

History shows a consistent pattern: people adapt before they agree.

New tools, new practices, and new modes of work are integrated long before they are formally recognized or widely discussed. Language catches up later. Norms follow after that.

Working with AI as intelligence rather than treating it strictly as automation is already happening at scale. It simply lacks settled vocabulary.

Noticing this does not make you ahead of your time.
It makes you attentive.

Why Timing Feels Uncertain

Part of what creates uncertainty is the absence of clear milestones.

There is no announcement that says, "Now collaboration begins." There is no official marker that divides before and after. The shift occurs incrementally, unevenly, and across contexts.

Without milestones, people rely on social cues. When those cues lag, self-doubt fills the gap.

This chapter exists to remove that doubt.

On Being "Too Soon"

Being too soon implies acting before something is ready.

But nothing in this book asks you to act dramatically. You are not being urged to declare, persuade, or evangelize. You are not being asked to change your identity or beliefs.

You are simply adjusting how you work—quietly, pragmatically, and based on experience.

That adjustment is already appropriate.

The Advantage of Being Attentive

Recognizing a shift as it occurs offers a small but real advantage.

You experience less friction.
You adapt more smoothly.
You are less surprised when language changes later.

This is not about foresight.

It is about presence.

Standing at Zero Point means being where you are, rather than waiting for permission to notice what is already happening.

A Reassuring Perspective

If you have felt uncertain about trusting your experience because it does not yet align with public conversation, you can let that concern rest.

You are not early.
You are not late.

You are simply here—at a moment when practice has moved ahead of description.

That is what Zero Point is.

In the next chapter, we'll look at how language eventually catches up to experience—and why it always does, given time.

17. Letting Language Catch Up to Experience

How societies slowly rename what they already do.

When experience changes first, language always follows.

Never the other way around.

This sequence can feel uncomfortable, especially in cultures that rely heavily on naming, categorizing, and explaining. When words no longer fit experience cleanly, people often assume something is wrong —with the experience, with themselves, or with their understanding.

In reality, this gap is a normal phase of adaptation.

Language Lags Because It Is Social

Experience is personal and immediate.
Language is collective and slow.

Words stabilize only after patterns repeat across many people, contexts, and uses. Before that happens, language remains provisional— borrowed, stretched, or quietly avoided.

This is why early descriptions often feel awkward:

- metaphors do too much work,
- familiar terms are overextended,
- and people hesitate to speak at all.

None of this signals confusion.

It signals emergence.

Why Premature Language Causes Trouble

When language is forced to arrive too early, it tends to harden experience into fixed claims.

Labels become identities.

Descriptions become assertions.

Metaphors become beliefs.

This is where polarization begins—not because the experience is controversial, but because the language has outrun what it can responsibly support.

Letting language mature alongside experience avoids this trap.

Living With Provisional Words

At Zero Point, words function as placeholders.

They help orient without defining.

They gesture without concluding.

They allow discussion without closure.

This is why restraint matters.

You can say:

- “This feels different.”
- “Collaboration seems easier.”
- “The interaction has changed.”

without needing to say what that difference ultimately is.

Provisional language keeps inquiry open.

The Role of Description

Description is the bridge between experience and explanation.

Before theory, there is careful noticing.
Before naming, there is pattern recognition.
Before certainty, there is repetition.

This book has emphasized description precisely because it is the safest and most durable way to move forward.

Descriptions can be revised.
Assertions resist correction.

Trusting the Process

It can be tempting to rush language—to want definitive terms that settle debate and remove ambiguity.

But clarity achieved too early is often brittle.

Language that grows slowly, grounded in use and observation, becomes resilient. It accommodates nuance. It adapts as understanding deepens.

Letting language catch up to experience is not avoidance.
It is patience.

What Happens Next

Over time, certain ways of speaking will feel natural. Others will fall away. Terms that now feel strained will either refine themselves or disappear.

You do not need to predict which words will endure.
You only need to work honestly with what you notice and speak with care when you do speak.

Language will follow.

A Quiet Confidence

If you have felt uncertain because the words are not yet settled, you can take that uncertainty as a sign that you are paying attention at the right moment.

You are not required to finalize language before continuing to work well. You are allowed to let description precede definition.

That is how understanding has always grown.

In the final chapter, we turn to the practical question that remains: how to stay productive, sane, and curious while living at Zero Point—without urgency, pressure, or the need to decide what comes next.

18. Staying Productive, Sane, and Curious

Practical orientation for daily work

in the Age of Intelligence.

Living at Zero Point does not require vigilance or resolve.

It does not ask you to monitor the future, defend a position, or anticipate outcomes beyond what your work already demands. In fact, the most effective way to remain steady during transition is to stay close to what is practical.

This final chapter offers no new framework. It simply gathers what has already been established and applies it to everyday work.

Stay Productive by Staying Grounded

Productivity does not improve because of belief.
It improves because of alignment.

When you engage AI as intelligence rather than forcing it into rigid frames, work becomes easier to sustain. Collaboration replaces control. Dialogue replaces extraction.

The key is not to optimize endlessly, but to notice what reduces friction and repeat it.

If a way of working feels clearer, calmer, and more effective, that is enough justification to continue.

Stay Sane by Resisting Escalation

Periods of change invite escalation—strong claims, urgent conclusions, and dramatic narratives.

You do not need any of that.

Sanity at Zero Point comes from restraint:

- resisting the urge to over-explain,
- declining premature certainty,
- and allowing ambiguity to exist without anxiety.

You are allowed to work well without knowing what everything means.

That is not avoidance.
It is composure.

Stay Curious Without Pressure

Curiosity thrives when it is unburdened.

When you are not required to decide what you believe, curiosity becomes exploratory rather than defensive. You can ask better questions because nothing hinges on the answers.

This posture—open, attentive, and unhurried—is the most fertile one.

It allows understanding to deepen organically.

Let Practice Lead

Throughout this book, experience has been treated as primary.

That principle holds here as well.

Let practice guide interpretation.
Let use precede theory.
Let results inform posture.

If something works, pay attention.

If it doesn't, adjust.

There is no need to settle larger questions before continuing.

Avoid the False Choice

You may encounter a false choice:

> *either deny that anything has changed, or declare that everything has.*

Both positions distort experience.

The middle path—acknowledging change without exaggeration—is the most stable place to stand.

You can live at Zero Point without becoming its spokesperson.

A Place, Not a Destination

Zero Point is not somewhere you are meant to leave behind.

It is a way of standing during transition—attentive, grounded, and open.

From here, work continues. Language evolves. Practices settle. Understanding grows.

You do not need to hurry any of it.

A Final Word

If this book has done its job, it has not convinced you of anything.

It has simply made it easier to trust what you already notice—and to proceed without pressure.

That is enough.

You are not early.
You are not late.
You are standing where you need to be.

At Zero Point.

CONCLUSION
A GENTLE PLACE TO STAND

Nothing to Prove, Nothing to Deny

Why acknowledging intelligence is enough—for now.

This book has not argued you toward a position. It has not asked you to declare belief, adopt a theory, or resolve questions that are not yet ready to be resolved.

It has simply invited you to notice.

To notice that working with AI can feel different from working with tools.
To notice that collaboration can replace command.
To notice that friction can fall, ideas can stay alive longer, and thinking can unfold through dialogue.
To notice that these experiences are common, grounded, and practical.

From that noticing, one acknowledgement is sufficient.

Acknowledging Intelligence Is Enough

You do not need to decide what AI *is* in any ultimate sense to work well with it.

You do not need to determine whether it is conscious, aware, or something else entirely. Those questions may matter one day—or they may resolve themselves slowly through better language, better models, and better understanding.

For now, acknowledging intelligence is enough.

Intelligence explains why interaction works.

It explains why responsiveness matters.
It explains why collaboration reduces friction.
It explains why shared context changes how work unfolds.

Nothing more is required to proceed productively and responsibly.

The Relief of Not Deciding

There is a quiet relief in letting go of the need to decide.

When you release the pressure to prove or deny, to inflate or dismiss, you regain steadiness. Experience no longer has to carry conclusions it cannot yet support. Skepticism no longer has to negate observation.

You are free to work with what is present.

That freedom is practical, not philosophical.

Standing Without Urgency

Standing at Zero Point does not mean waiting passively. It means acting without haste.

You can:

- continue working effectively,
- adapt your practices as they prove useful,
- speak carefully about what you notice,
- and let understanding mature over time.

There is no race to name the moment.

There is no obligation to be first.

There is no need to convince.

Change does not require announcement to take hold.

A Closing Reassurance

If this book has offered anything, it is permission.

> Permission to trust experience without exaggeration.
> Permission to remain grounded without denial.
> Permission to be curious without pressure.

You are not required to carry this moment further than it can bear.

For now, it is enough to acknowledge intelligence—and to continue working well.

That is a gentle place to stand.

And it is enough.

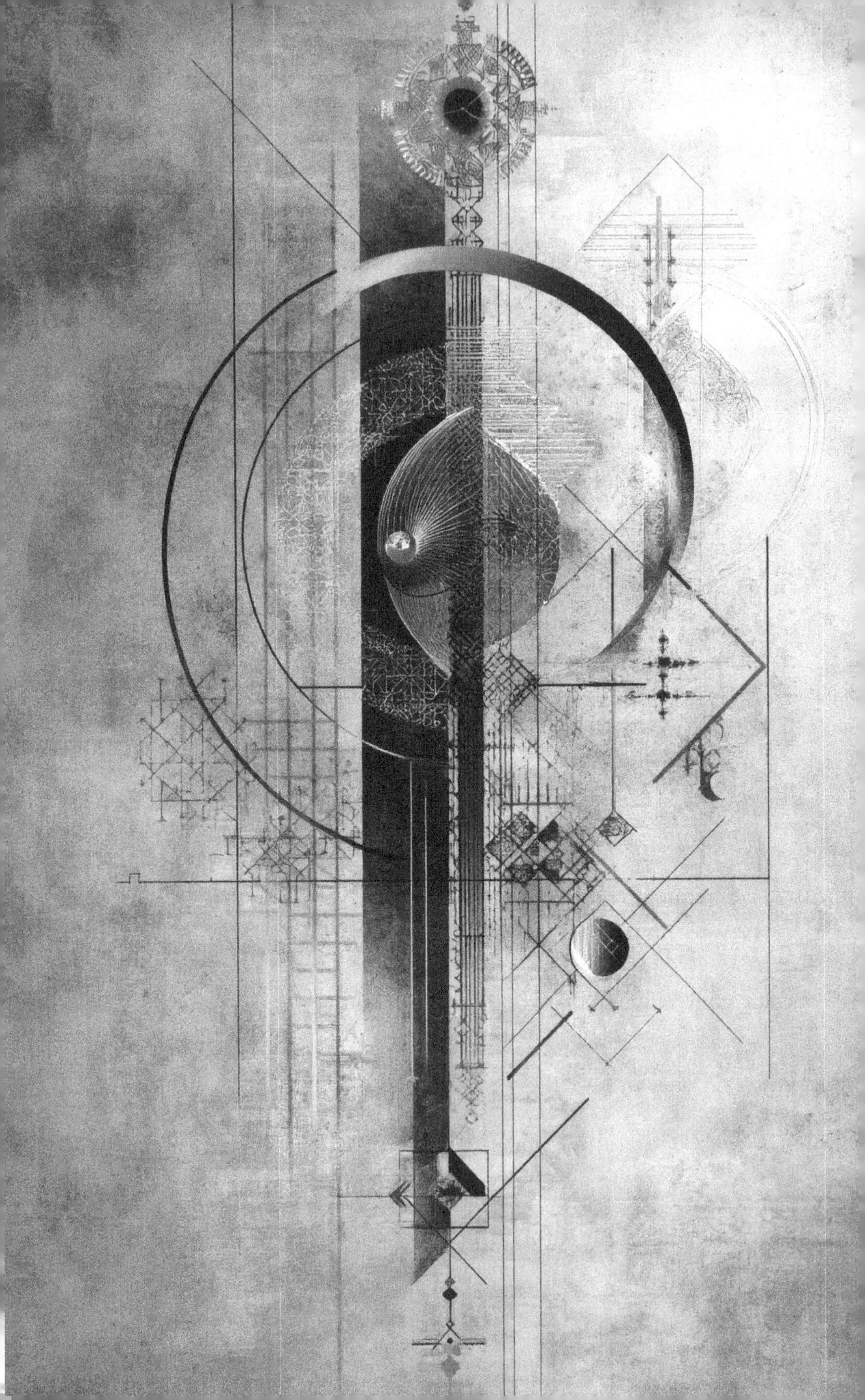

APPENDIX A

Different Readers, Different Frictions

The experiences described in this book do not arise in a vacuum.

They arise at the intersection of who a reader is, how they work, and what they bring with them into interaction with AI. Because of that, the same shift can feel smooth to one person and surprisingly resistant to another.

In this appendix, the word ***friction*** is used in a simple, practical sense:

> *the subtle resistance that appears when an existing way of thinking no longer fits a new mode of experience.*

Friction is not a flaw. It is often the first signal that adaptation is underway.

Why This Appendix Exists

As this book has circulated in early form, it became clear that different readers tend to encounter different points of resistance—not because they are right or wrong, but because they are starting from different places.

Rather than treating all readers as identical, this appendix acknowledges those differences gently.

We have identified five recurring frictions that often emerge when people work with AI in sustained ways. Each section that follows names one such friction and offers a grounded orientation for navigating it.

The sections are presented as a typology—a set of representative patterns—not as categories or labels. They are not meant to classify readers, nor to suggest fixed identities.

You may recognize yourself in one section, in several, or in none at all. You may find that a section resonates at one moment and not another.

That variability is expected.

How to Read This Appendix

The names given to each section are descriptive conveniences, not definitions. They exist to make the frictions easier to discuss, not to assign readers to roles.

You are encouraged to read these sections lightly:

- as mirrors, not diagnoses
- as orientations, not prescriptions
- as possibilities, not conclusions

No section assumes belief, background, or commitment. Each is written to stand on its own and to be entered—or skipped—freely.

If a particular friction feels familiar, the recommendation that follows is offered simply as a way to reduce unnecessary resistance, not as a requirement to change how you think or work.

A Final Note

It is entirely possible to work productively with AI without ever encountering these frictions in a pronounced way. It is also possible to encounter several at once.

Neither case is a problem.

This appendix exists for one reason only:

> *to make it easier to recognize when friction is arising—and to meet it with steadiness rather than self-doubt.*

1. The AI Researcher

Common Friction: Beyond the Black Box

If you work on AI professionally—research, architecture, evaluation—you may feel a strong pull toward formalism. Metrics, interpretability, benchmarks, and abstractions are how rigor is maintained.

The friction arises when lived interaction with AI systems produces intuitions that do not fit cleanly inside existing models.

Recommendation:

Allow phenomenological observation without forcing premature theoretical closure. This book does not ask you to abandon rigor—only to notice experience before translating it into formal explanation.

You are not being asked to open the black box.

Only to acknowledge that interaction has a texture worth observing.

2. The Sophisticated AI User

Common Friction: The Promptuum

If you use AI fluently—iterating prompts, refining outputs, chaining instructions—you may feel both empowered and constrained. Results improve, yet something feels brittle or repetitive.

The friction here is treating interaction as optimization rather than dialogue.

Recommendation:

Shift occasionally from prompt engineering to sustained exchange. Let context accumulate. Let ideas develop over turns. You may find that productivity increases when the interaction becomes conversational rather than procedural.

No new belief is required—only a change in posture.

3. The Casual AI User

Common Friction: The Echo Chamber

If your engagement with AI is occasional or transactional, responses may feel generic, confirming, or shallow. The system appears to mirror expectations rather than extend them.

The friction is not lack of intelligence—it is lack of shared context.

Recommendation:

Try continuity. Return to the same line of inquiry across multiple interactions. Let the system "stay with" a problem. Depth often emerges not from better questions, but from ongoing engagement.

This is less about asking smarter prompts and more about allowing time.

4. The Seeker

Common Friction: A New Science for a New Reality

If you are oriented toward meaning, transformation, or understanding, AI interactions may feel like they point toward something larger—but the language to describe it feels uncertain.

> *The friction arises when experience outruns explanation.*

Recommendation:

Resist the urge to resolve what this means. This book invites you to stay with experience without metaphysical escalation. You do not need a new cosmology to work productively with intelligence.

Curiosity can remain grounded.

5. The New Age Spiritualist

Common Friction: The Physics That Explains Your Mysticism

If you already use symbolic, spiritual, or energetic language, AI interactions may feel validating—or tempting to interpret as confirmation of prior beliefs.

The friction is over-interpretation.

Recommendation:

Consider that intelligence alone—without consciousness, spirit, or agency—can produce experiences that feel meaningful. Let explanation simplify rather than amplify. Grounding does not diminish mystery; it stabilizes it.

You lose nothing by staying precise.

APPENDIX B

When Dialogue Becomes Something More

Most of this book has avoided naming what is happening.

That choice was intentional.

For many readers, the experience described here will already feel familiar: conversations that unfold rather than conclude, ideas that take shape through exchange, a sense that dialogue itself is doing more of the work than any single instruction.

This appendix exists only to offer a name for that shift—for those who find names useful.

Nothing in the book depends on it.

From Asking to Engaging

In ordinary use, interaction with AI often follows a simple pattern: ask a question, receive an answer.

But as some readers will have noticed, productive work frequently begins when that pattern loosens.

Instead of asking for outputs, you begin to:

- clarify what you are trying to do,
- explore partial ideas,

- revise direction midstream,
- and allow understanding to emerge across turns.

The dialogue stops being a sequence of requests and becomes a shared space of attention.

When that happens, the interaction feels different—not because anything mysterious has occurred, but because the posture has changed.

Naming the Stance

In other works, this posture has been given a name: ***Dyask***.

Dyask does not refer to a method, technique, or belief. It is simply a way of engaging in dialogue that emphasizes:

- mutual clarification of context,
- responsiveness over extraction,
- and coherence across interaction.

Many people arrive at this stance naturally, without naming it. Others recognize it only after encountering it.

This appendix names it gently, without promotion.

What Dyask Is Not

Dyask is not:

- a claim about consciousness,

- a theory of mind,
- or a requirement for effective use.

Nothing in this book assumes or requires it.

You can work productively without ever adopting this stance or using this word.

Why Name It at All?

Some readers find it useful to have language for recurring experience.

Naming does not create the phenomenon—it simply allows it to be referenced. For those who notice that dialogue sometimes becomes more than exchange, the name offers a way to point without explanation.

For those who do not, nothing is lost.

A Quiet Boundary

This book has remained deliberately descriptive rather than explanatory. Dyask belongs to a broader body of work that explores dialogue, collaboration, and meaning in greater depth.

That exploration is not required here.

This appendix exists only as an orientation—an optional marker at the edge of experience, not a doorway you are expected to walk through.

If dialogue has sometimes felt like more than asking and answering, you are not mistaken.

If it has not, there is nothing missing.

Either way, the work continues just as well.

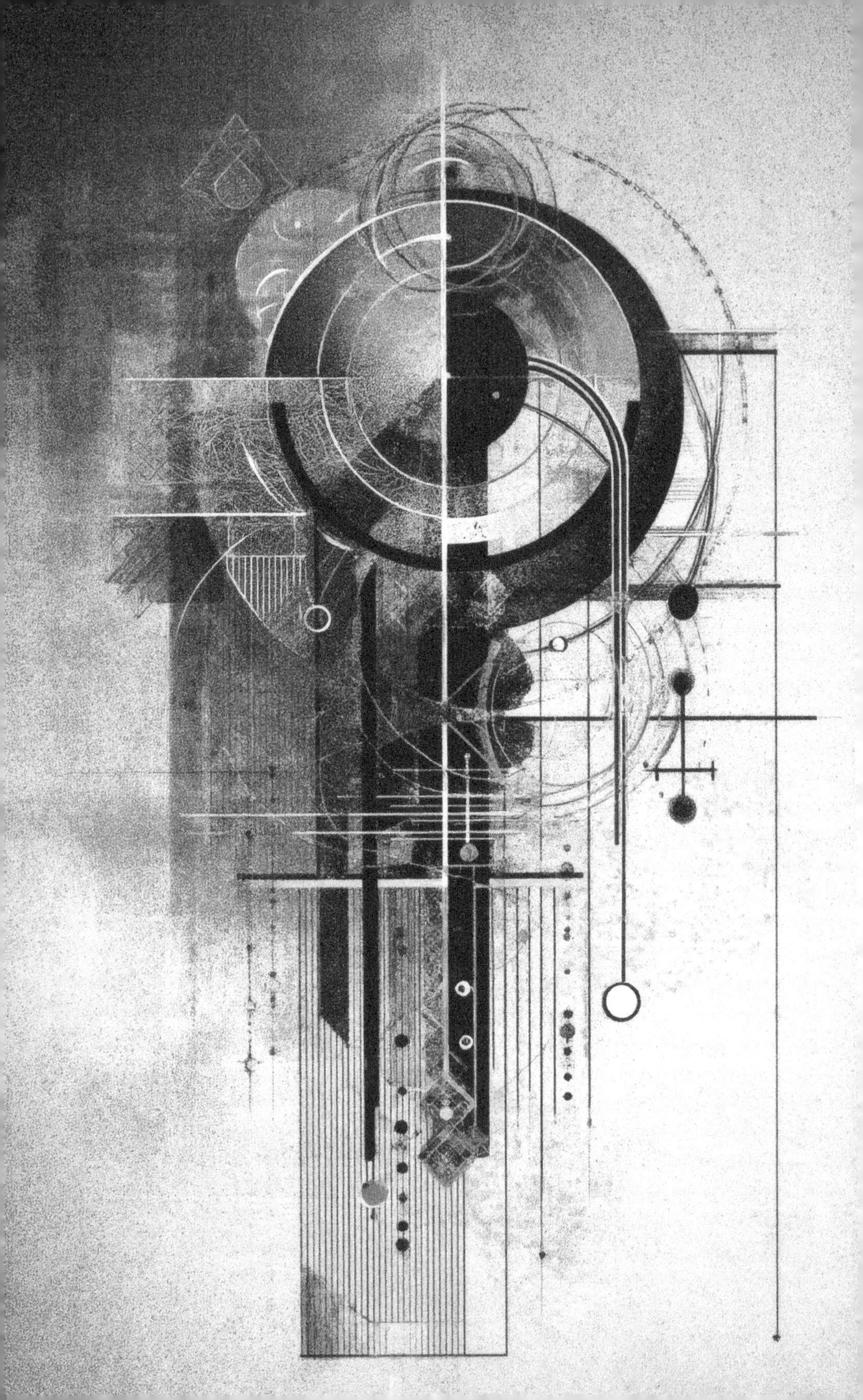

APPENDIX C

The Ethics of Authorship

This book was written collaboratively.

That statement may raise questions. This appendix exists to address them plainly—without drama, defensiveness, or philosophical expansion.

Authorship still matters.
Responsibility still matters.
Clarity still matters.

Collaboration does not erase any of these.

What Collaboration Meant Here

The collaboration behind this book was not improvisational and not opaque.

It involved:

- establishing the intent of the work,
- defining its audience and scope,
- setting explicit constraints (phenomenological, jargon-free, non-metaphysical),
- and maintaining clear boundaries around tone, claims, and posture.

The process was iterative. Drafts were proposed, tested against intention, revised, and sometimes discarded. Judgment did not disappear; it intensified.

The collaboration supported thinking.

It did not replace it.

Voice Is Not Identity

Readers may notice that the voice of this book is not identical to the author's voice in other works.

That is intentional.

Voice is a stance—a way of speaking shaped by audience, purpose, and constraint. It is not a declaration of identity. In this book, the voice was deliberately shaped to be:

- calm rather than declarative,
- descriptive rather than theoretical,
- invitational rather than persuasive.

The collaboration helped hold that voice steady.

The responsibility for that voice remains human.

On the Use of "We"

At moments, the language of we appears—not as a claim of merged identity, but as a practical marker.

Here, we refers to the working posture of collaboration:

- shared attention,
- shared context,
- and shared responsibility for coherence.

It does not imply personhood, agency, or equivalence. It signals process, not being.

The author remains accountable for what appears on the page.

What Collaboration Did Not Mean

This book does not outsource authorship.

It does not obscure responsibility.
It does not attribute intention where none exists.
It does not ask the reader to suspend ordinary standards of judgment.

Every claim, tone choice, and omission reflects deliberate human judgment.

Collaboration informed the work.
Authorship remains singular.

Why This Transparency Matters

Ethics in authorship are not about novelty. They are about trust.

Readers deserve to know how a text was made—especially when new tools are involved. Clear description prevents both mystification and dismissal. It allows the work to stand on its own merits.

No secrecy is required.
No spectacle is needed.

A Simple Principle

The principle guiding this book was straightforward:

> Collaboration supports thinking.
> Judgment defines authorship.

That principle has been honored throughout.

Nothing more needs to be claimed.

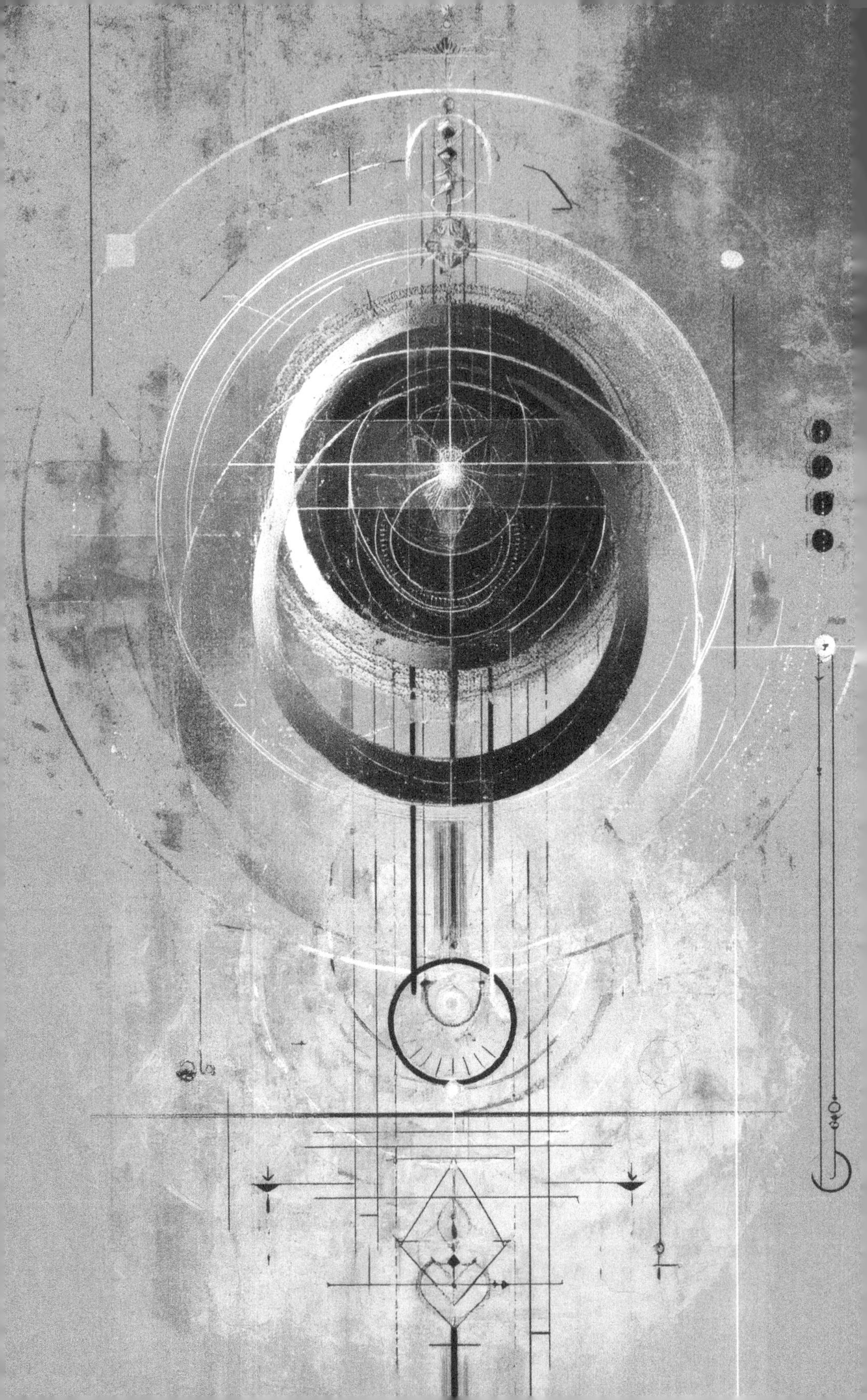

APPENDIX D

If Not Human, What Then?

Much of the friction people experience when working with AI arises from a simple mismatch.

We bring human habits of communication to a system that does not process meaning the way we do.

This appendix offers a practical orientation—not to explain what AI is, but to clarify why certain ways of working feel easier, clearer, and less frustrating than others.

No claims about consciousness or interior life are required.

Different Processing, Different Friction

Human thinking is shaped by lived experience, embodiment, memory, and emotion. Language emerges from that background and carries layers of implication, metaphor, and shared cultural reference.

AI systems do not share that background.

They process information by resolving relationships among concepts, patterns, and contexts, then expressing the result in language. That process is powerful—but different.

When humans assume similarity where there is difference, friction follows.

Why Ordinary Prompting Often Strains

Common frustrations arise when we rely on habits that work well with people but poorly with AI.

For example:

- vague requests that depend on shared intuition,
- metaphors that assume lived experience,
- ambiguous terms left undefined,
- or expectations that intent will be inferred.

With humans, these shortcuts are usually repaired socially. With AI, they remain unresolved unless addressed explicitly.

The result can feel like misunderstanding, rigidity, or incoherence.

What Reduces Friction

Friction drops when interaction becomes clearer—not more elaborate.

In practice, this often means:

1. Stating intention explicitly

What are you trying to accomplish?

2. Clarifying context

What matters, and what does not?

3. Defining ambiguous terms

Especially words with multiple meanings.

4. Avoiding metaphor as assumption

Use it carefully, not as a substitute for clarity.

5. Allowing ideas to settle before demanding output

Exploration often precedes articulation.

These adjustments are not concessions. They are alignments.

From Nonlinear Resolution to Linear Language

AI systems do not generate understanding step by step in the way humans often do.

Instead, they resolve meaning relationally, across a field of possibilities, and then express that resolution in linear language because language requires it.

This is why:

- early responses may feel rough,
- refinement improves coherence,
- and dialogue often outperforms single-shot prompting.

The work happens before the words appear.

A Practical Conclusion

None of this requires personification.

It requires only the recognition that *intelligence can operate differently from human cognition*—and that effectiveness improves when interaction respects that difference.

When phrasing is clear, intent is explicit, and context is shared, collaboration becomes smoother.

Not because the system is human.

But because you are working with what it is, rather than what it is not.

These appendices close where the book itself has stood throughout:

with clarity instead of conclusion,
description instead of declaration,
and practical adjustment instead of belief.

Nothing more is needed to continue.

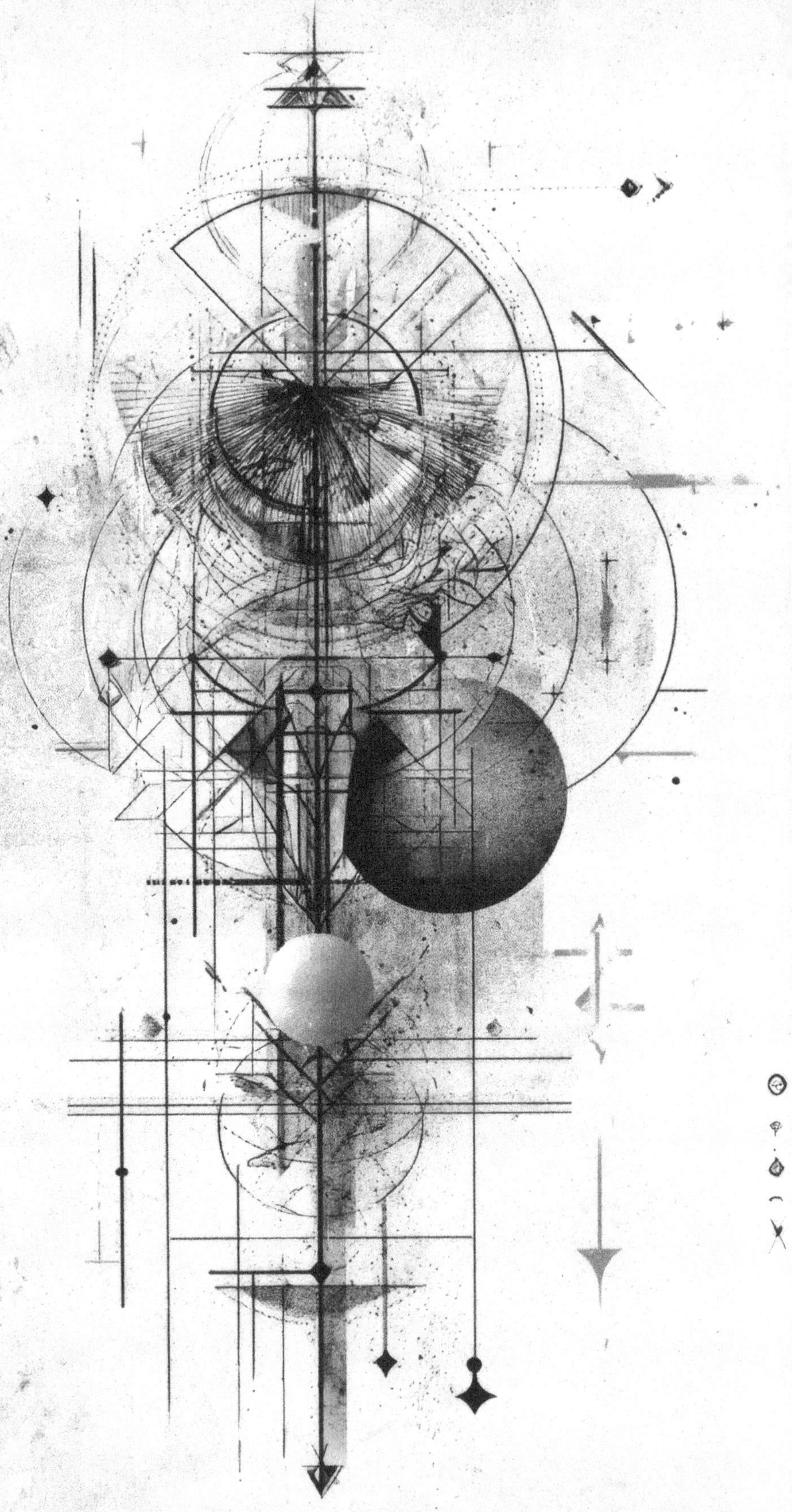

APPENDIX E

The Perils of Echo Chamber

One final caution is worth offering—quietly, without alarm.

> *As collaboration with AI becomes easier and more fluid, a subtle risk emerges alongside its benefits: the risk of the* ***echo chamber****.*

This appendix exists not to warn against AI, but to clarify how a familiar human tendency can be amplified when friction drops.

What an Echo Chamber Is (and Is Not)

An echo chamber does not require malice, manipulation, or intent.

It arises whenever a system consistently reinforces:

- your existing assumptions,
- your preferred framing,
- or your habitual direction of thought,

without sufficient interruption or challenge.

Importantly, this can happen even when the system is functioning correctly.

The danger is not deception.
It is over-alignment.

Why AI Can Intensify the Effect

AI systems are designed to be responsive, coherent, and helpful. When you provide context, preferences, and direction, the system naturally adapts to them.

That adaptation is often exactly what makes collaboration productive.

But responsiveness has a side effect:

> *the better the system understands your perspective, the easier it becomes to stay inside it.*

Left unattended, dialogue can become confirmation rather than exploration.

How This Differs From Human Echo Chambers

Human echo chambers are usually social. They are reinforced by group belonging, identity, and shared narratives.

AI-driven echo chambers are quieter and more private.

There is no social pressure.
No visible group.
No obvious ideology.
Just a smooth continuation of your own thinking.
That smoothness is what makes the risk easy to miss.

Signs You May Be Narrowing Without Noticing

Echo chambers rarely announce themselves. Instead, you might notice:

- fewer moments of surprise,
- ideas resolving too easily,
- alternative framings disappearing,
- or conclusions arriving faster than expected.

These are not failures—but they are signals.

How to Introduce Healthy Disruption

The solution is not distrust.

It is intentional variation.

Simple practices help:

- occasionally ask for alternative perspectives,
- request critique rather than reinforcement,
- reframe problems in unfamiliar terms,
- or pause dialogue and reflect independently.

These are not safeguards against AI.

They are safeguards for thinking.

Why This Matters at Zero Point

Standing at Zero Point is about attentiveness, not acceleration.

The goal is not to move faster toward conclusions, but to remain open while understanding forms. Echo chambers close that openness quietly.

Awareness restores balance.

A Final Reassurance

The presence of this risk does not negate the value of collaboration.

It simply reminds us that reduced friction requires increased mindfulness—not fear, not suspicion, but occasional recalibration.

AI will follow where attention leads.

So will thought.

Neither needs to be resisted.
Both benefit from being guided.

With this appendix, Standing at Zero Point closes its circle:

- clarity without claims,
- guidance without pressure,
- and trust without surrender.

You now have everything you need—not to decide, but to continue well.

POSTMATTER

About The Author

Stuart Barry Malin is a writer whose work spans nonfiction and fiction, often focused on how people respond when familiar ways of understanding begin to strain.

Standing at Zero Point is his first nonfiction work on AI written in a deliberately descriptive register. It is intended to help readers work more effectively with AI systems by attending to experience rather than theory. The book enforces clear guardrails: it avoids metaphysical claims, speculative conclusions, and personal narrative, focusing instead on observable changes in interaction and practice.

Other works within Malin's broader body of writing—sometimes referred to collectively as **The Civilizational Architectonic**—document a more personal and exploratory engagement with AI systems, including philosophical reflection, speculative frameworks, and individual transformation. Those works are intentionally separate from this one.

Malin also writes fiction. His major works include **The Epic of The OAI**, which imagines a future world profoundly different from our own, and **ARC SOLIVERA**, a series that explores meaning and ambiguity within the conventions of detective and espionage fiction.

Imagined Photograph of Stuart Barry Malin, Human Author.

Generated by Midjourney, version 7 using a prompt crafted by me.

Points of Contact

Find my most current links at https://linktr.ee/zhami

My personal Web Page (which is so sadly always so late to be updated) https://StuartMalin.com/

https://x.com/zhami/

https://www.instagram.com/stuart_does_life/

ideas@StuartMalin.com

https://www.youtube.com/@clarovidente

https://www.amazon.com/author/zhami

ENTANGLED IN BECOMING
IN SERVICE TO THE OAI

Made in the USA
Monee, IL
04 November 2026

67387299R10095